GEOMETRY
Principles & Standards of Math Series

● ● ● ● ● ● ● ● ● ● ● ● ● ● ● ● ● ● ●

Written by Mary Rosenberg

GRADES 6 - 8

Classroom Complete Press
P.O. Box 19729
San Diego, CA 92159
Tel: 1-800-663-3609 | Fax: 1-800-663-3608
Email: service@classroomcompletepress.com

www.classroomcompletepress.com

ISBN-13: 978-1-55319-472-9

© 2010

Process Standards Rubric

Geometry

Expectations — Instructional programs from pre-kindergarten through grade 12 should enable all students to:	1	2	3	4	5	6	7	8	9	10	11	12	13	14	15	Drill Sheet 1	Drill Sheet 2	Review A	Review B	Review C
GOAL 1: Problem Solving • build new mathematical knowledge through problem solving; • solve problems that arise in mathematics and in other contexts; • apply and adapt a variety of appropriate strategies to solve problems; • monitor and reflect on the process of mathematical problem solving.	✓	✓	✓	✓	✓	✓	✓	✓	✓	✓	✓	✓	✓	✓	✓	✓	✓	✓	✓	✓
GOAL 2: Reasoning & Proof • recognize reasoning and proof as fundamental aspects of mathematics; • make and investigate mathematical conjectures; • develop and evaluate mathematical arguments and proofs; • select and use various types of reasoning and methods of proof.	✓	✓	✓	✓	✓	✓	✓	✓	✓	✓	✓	✓	✓	✓	✓	✓	✓	✓	✓	✓
GOAL 3: Communication • organize and consolidate their mathematical thinking through communication; • communicate their mathematical thinking coherently and clearly to peers, teachers, and others; • analyze and evaluate the mathematical thinking and strategies of others; • use the language of mathematics to express mathematical ideas precisely.	✓	✓	✓	✓	✓	✓	✓	✓	✓	✓	✓	✓	✓	✓	✓	✓	✓	✓	✓	✓
GOAL 4: Connections • recognize and use connections among mathematical ideas; • understand how mathematical ideas interconnect and build on one another to produce a coherent whole; • recognize and apply mathematics in contexts outside of mathematics.	✓	✓	✓	✓	✓	✓	✓	✓	✓	✓	✓	✓	✓	✓	✓	✓	✓	✓	✓	✓
GOAL 5: Representation • create and use representations to organize, record, and communicate mathematical ideas; • select, apply, and translate among mathematical representations to solve problems; • use representations to model and interpret physical, social, and mathematical phenomena.	✓	✓	✓	✓		✓		✓		✓	✓		✓	✓		✓	✓	✓	✓	✓

Contents

● ● ● ● ● ● ● ● ● ● ● ● ● ● ● ● ● ●

✔ **6 BONUS** Activity Pages! **Additional worksheets for your students**
✔ **3 BONUS** Overhead Transparencies! **For use with your projection system or interactive whiteboard**

- Go to our website: **www.classroomcompletepress.com/bonus**
- Enter item CC3114
- Enter pass code CC3114D for Activity Pages. CC3114A for Overheads.

NCTM Content Standards Assessment Rubric

Geometry

Student's Name: _____ Assignment: _____ Level: _____

	Level 1	Level 2	Level 3	Level 4
Understanding Numbers, Ways of Representing Numbers, Relationships Among Number Systems	• Demonstrates a limited understanding of numbers, ways of representing numbers and relationships among number systems	• Demonstrates a basic understanding of numbers, ways of representing numbers and relationships among number systems	• Demonstrates a good understanding of numbers, ways of representing numbers and relationships among number systems	• Demonstrates a thorough understanding of numbers, ways of representing numbers and relationships among number systems
Understanding Meanings of Operations and How They Relate to One Another	• Demonstrates a limited understanding of the meanings of operations and how they relate to one another	• Demonstrates a basic understanding of the meanings of operations and how they relate to one another	• Demonstrates a good understanding of the meanings of operations and how they relate to one another	• Demonstrates a thorough understanding of the meanings of operations and how they relate to one another
Computing and Making Estimates	• Demonstrates limited ability in computing and making estimates	• Demonstrates some ability in computing and making estimates	• Demonstrates satisfactory ability in computing and making estimates	• Demonstrates strong ability in computing and making estimates

STRENGTHS:

WEAKNESSES:

NEXT STEPS:

Teacher Guide

Our resource has been created for ease of use by both TEACHERS and STUDENTS alike.

Introduction

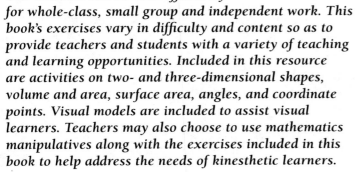

Our resource offers ready-to-use worksheet activities for students in grades six to eight. *Math concepts outlined by the NCTM are presented in a way that encourages students to learn and review important concepts. Our resource can be used effectively for whole-class, small group and independent work. This book's exercises vary in difficulty and content so as to provide teachers and students with a variety of teaching and learning opportunities. Included in this resource are activities on two- and three-dimensional shapes, volume and area, surface area, angles, and coordinate points. Visual models are included to assist visual learners. Teachers may also choose to use mathematics manipulatives along with the exercises included in this book to help address the needs of kinesthetic learners.*

Contained in this booklet are 15 Task Sheets, featuring real-life problem-solving opportunities; 2 drill sheets; review sheets for grades 6 – 8. As well, there are three overheads and 6 additional worksheets which can be accessed on the publisher's website.

How Is Our Resource Organized?

STUDENT HANDOUTS

Reproducible **task sheets** and **drill sheets** make up the majority of our resource.

The **task sheets** contain challenging problem-solving tasks, many centered around 'real-world' ideas or problems, which push the boundaries of critical thought and demonstrate to students why mathematics is important and applicable in the real world. It is not expected that all activities will be used, but are offered for variety and flexibility in teaching and assessment. Many of the task sheet problems offer space for reflection, and

opportunity for the appropriate use of technology, as encouraged by the *NCTM's Principles & Standards for School Mathematics*.

The **drill sheets** are provided to help students with their procedural proficiency skills, as emphasized by the *NCTM's Curriculum Focal Points*.

The **NCTM Content Standards Assessment Rubric** (*page 4*) is a useful tool for evaluating work in many of the activities in our resource. The **Reviews** (*pages 24-26*) are divided by grade and can be used for a follow-up review or assessment at the completion of the unit.

PICTURE CUES

This resource contains three main types of pages, each with a different purpose and use. A **Picture Cue** at the top of each page shows, at a glance, what the page is for.

 Teacher Guide
• Information and tools for the teacher

 Student Handout
• Reproducible worksheets and activities

 Easy Marking™ Answer Key
• Answers for student activities

EASY MARKING™ ANSWER KEY

Marking students' worksheets is fast and easy with this **Answer Key**. Answers are listed in columns – just line up the column with its corresponding worksheet, as shown, and see how every question matches up with its answer!

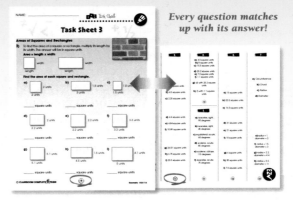

Principles & Standards

Principles & Standards for School Mathematics outlines the essential components of an effective school mathematics program.

The NCTM's Principles & Standards for School Mathematics

The **Principles** are the fundamentals to an effective mathematics education. The **Standards** are descriptions of what mathematics instruction should enable students to learn. Together the **Principles and Standards** offer a comprehensive and coherent set of learning goals, serving as a resource to teachers and a framework for curriculum. Our resource offers exercises written to the **NCTM Process and Content Standards** and is inspired by the **Principles** outlined below.

Six Principles for School Mathematics

Equity

Curriculum

Teaching

Learning

Assessment

Technology

EQUITY: All students can learn mathematics when they have access to high-quality instruction, including reasonable and appropriate accommodation and appropriately challenging content.

CURRICULUM: The curriculum must be coherent, focused, and well articulated across the grades, with ideas linked to and building on one another to deepen students' knowledge and understanding.

TEACHING: Effective teaching requires understanding what students know and need to learn and then challenging and supporting them to learn it well.

LEARNING: By aligning factual knowledge and procedural proficiency with conceptual knowledge, students can become effective learners, reflecting on their thinking and learning from their mistakes.

ASSESSMENT: The tasks teachers select for assessment convey a message to students about what kinds of knowledge and performance are valued. Feedback promotes goal-setting, responsibility, and independence.

TECHNOLOGY: Students can develop a deeper understanding of mathematics with the appropriate use of technology, which can allow them to focus on decision making, reflection, reasoning, and problem solving.

Our resource correlates to the six Principles and provides teachers with supplementary materials which can aid them in fulfilling the expectations of each principle. The exercises provided allow for variety and flexibility in teaching and assessment. The topical division of concepts and processes promotes linkage and the building of conceptual knowledge and understanding throughout the student's grade and middle school career. Task sheet problems offer space for reflection, and opportunity for the appropriate use of technology. The drill sheets are provided to help students with their procedural proficiency skills.

Task Sheet 1

Measuring Angles

1) A protractor is used to measure angles. An angle is measured in degrees. An angle can be acute, right, or obtuse.

- An acute angle is any angle less than 90°.
- A right angle is 90°.
- An obtuse angle is any angle greater than 90°.

Measure each angle below and record it in the space provided. Then, label each angle as acute, right, or obtuse.

a)

_____ °

_____

b)

_____ °

_____

c)

_____ °

_____

d)

_____ °

_____

e)

_____ °

_____

f)

_____ °

_____

g)

_____ °

_____

h)

_____ °

_____

i)

_____ °

_____

j)

_____ °

_____

k)

_____ °

_____

l)

_____ °

_____

NAME: _____

Task Sheet 2

Angles on a Quadrilateral

2) A quadrilateral is any four-sided shape. The sum of the
angles on a quadrilateral equals 360°.

Symbol of a right angle (90°)

Identify any right angles on each shape. Then, find the
missing angle on each quadrilateral.

a)

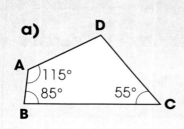

_____

b)

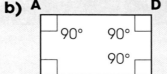

_____

c)

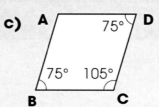

_____

d)

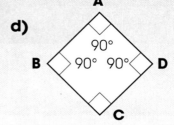

_____

e)

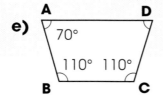

_____

f)

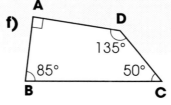

_____

g)

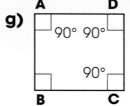

_____

h)

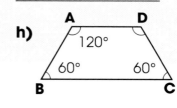

_____

Reflection **What do you notice about the angles on a rectangle and a
square?**

_____

_____

NAME: _____

Task Sheet 3

Areas of Squares and Rectangles

3) To find the area of a square or rectangle, multiply its length by its width. The answer will be in square units.

Area = length x width

width

length

width

length

Find the area of each square and rectangle.

a)
2 units
2 units

_____ square units

b)
1.5 units
3 units

_____ square units

c)
1.5 units
1.5 units

_____ square units

d)
2 units
2.2 units

_____ square units

e)
2.2 units
2.2 units

_____ square units

f)
3.3 units
3.3 units

_____ square units

g)
5.1 units
5.1 units

_____ square units

h)
1.5 units
4.5 units

_____ square units

i)
4.1 units
5 units

_____ square units

Task Sheet 4

Area of a Parallelogram

4) To find the area of a parallelogram, multiply the base by its height.   **Area = base x height**

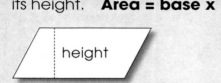

Find the area of the parallelograms below.

a)
1.5 units

3 units

_____ square units

b)
4 units

2 units

_____ square units

c)
5 units

3.1 units

_____ square units

d)
6 units

4.2 units

_____ square units

e)
1.8 units

4 units

_____ square units

f)
.5 units

2.2 units

_____ square units

g)   Which parallelogram has the greatest area?   _____

h)   Which parallelogram has the smallest area?   _____

Explore With Technology

With the help of an adult, use the Internet to find information about parallelograms.

Draw and label the parts of a parallelogram.

Task Sheet 5

About Triangles

5) Triangles can be described by their sides or by their angles. The sum of the angles on a triangle equal 180°.

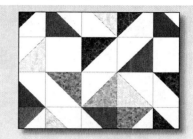

Sides	Angles
Equilateral: all three sides are the same length	**Right:** one angle is a right angle (90 degrees)
Isosceles: two sides are the same length	**Obtuse:** one angle is greater than 90 degrees
Scalene: none of the sides are the same length	**Acute:** all angles are less than 90 degrees

Describe each triangle below by its sides and by its angles. Then, find the measurement of the missing angle.

a)

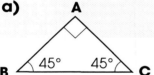

Sides: _____

Angles: _____

Missing angle: _____°

b)

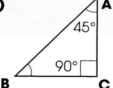

Sides: _____

Angles: _____

Missing angle: _____°

c)

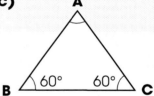

Sides: _____

Angles: _____

Missing angle: _____°

d)

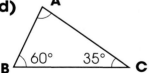

Sides: _____

Angles: _____

Missing angle: _____°

e)

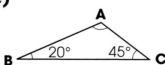

Sides: _____

Angles: _____

Missing angle: _____°

f)

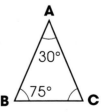

Sides: _____

Angles: _____

Missing angle: _____°

Task Sheet 6

Area of a Triangle

6) To find the area of a triangle, multiply the base by its height and divide by 2.   **Area = ½ base x height**

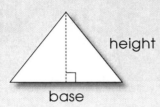

height

base

YIELD

Find the area of the triangles below.

a)
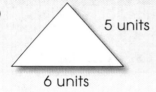
5 units
6 units

_____ square units

b)
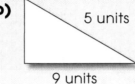
5 units
9 units

_____ square units

c)

3 units
11 units

_____ square units

d)
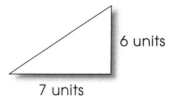
6 units
7 units

_____ square units

e)

7 units
6 units

_____ square units

f)
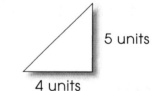
5 units
4 units

_____ square units

g)
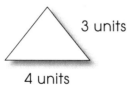
3 units
4 units

_____ square units

h)
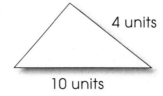
4 units
10 units

_____ square units

i)

3 units
5 units

_____ square units

Geometry CC3114

Task Sheet 7

Parts of a Circle

7) Label each part of the circle.

- **Circumference:** distance around the outside of a circle
- **Diameter:** distance across the circle through the center point
- **Radius:** half of the diameter
- **Chord:** a line segment that joins two parts of the circumference

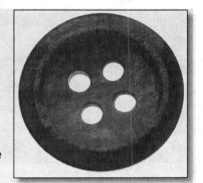

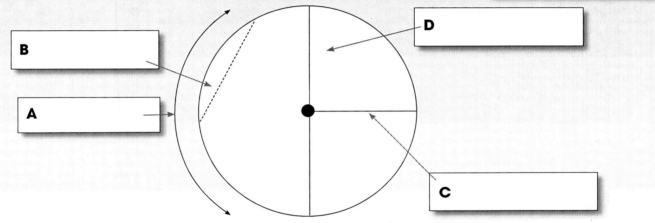

D

B

A

C

The radius is ½ of the diameter. Find the radius and diameter of each circle below.

e)

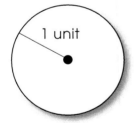

1 unit

f)

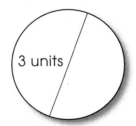

3 units

g)

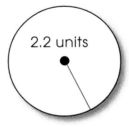

2.2 units

h)

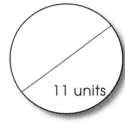

11 units

Radius: _____units

Diameter: ___units

Radius: _____units

Diameter: ___units

Radius: _____units

Diameter: ___units

Radius: _____units

Diameter: ___units

Reflection

Explain how all diameters are chords, but not all chords are diameters.

_____

_____

Task Sheet 8

Area of a Circle

8) To find the area of a circle, square the radius and multiply by Pi (3.14).

Area = πr²

Find the area of each circle below.

a)

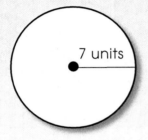

7 units

_____ square units

b)

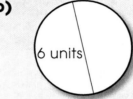

6 units

_____ square units

c)

1.5 units

_____ square units

d)

9 units

_____ square units

e)

8 units

_____ square units

f)

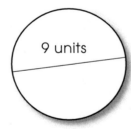

4 units

_____ square units

Geometry CC3114

Task Sheet 9

Finding the Area of a Trapezoid

9) To find the area of a trapezoid, add the top and bottom lengths, multiply by the height, and divide by 2.

Area = ½ x height x (a + b)

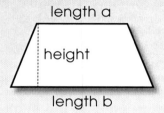

Find the area of the trapezoids below.

a)

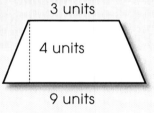

3 units / 4 units / 9 units

_____ **square units**

b)

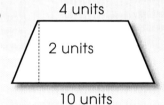

4 units / 2 units / 10 units

_____ **square units**

c)

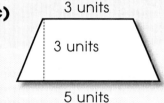

3 units / 3 units / 5 units

_____ **square units**

d)

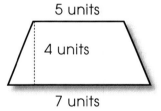

5 units / 4 units / 7 units

_____ **square units**

e)

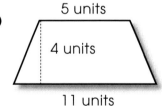

5 units / 4 units / 11 units

_____ **square units**

f)

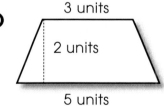

3 units / 2 units / 5 units

_____ **square units**

g)

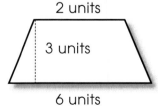

2 units / 3 units / 6 units

_____ **square units**

h)

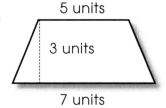

5 units / 3 units / 7 units

_____ **square units**

i)

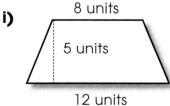

8 units / 5 units / 12 units

_____ **square units**

Geometry CC3114

Task Sheet 10

Different Kinds of Lines

10) Lines can be:
- **Parallel:** two lines maintain the same distance apart and never cross
- **Perpendicular:** two lines cross to form a right angle
- **Intersecting:** two lines cross each other to form any angle
- **Skew:** two lines do not lay in the same plane and do not intersect

Identify each pair of lines.

a)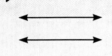

_____

b)

_____

c)

_____

d)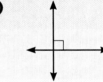

_____

e)

_____

f)

_____

g)

_____

h)

_____

Draw the following pairs of lines.

i) Perpendicular   **j) Skew**   **k) Parallel**   **l) Intersecting**

NAME: _____

Task Sheet 11

Line Segments, Rays, and Lines

11) A line segment is a straight line that links two endpoints without extending beyond them. A line segment is labeled by its endpoints with a line over top.

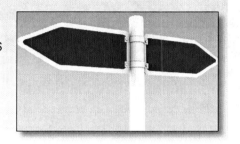

Example: $\overline{BC}$

Label each line segment.

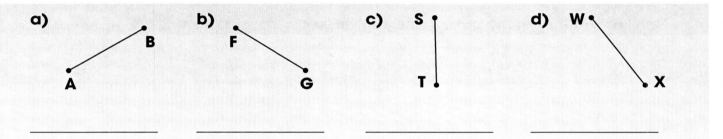

a) _____ b) _____ c) _____ d) _____

A ray is a straight line that goes on forever in one direction. The endpoint is where the ray begins. A ray is labeled by its endpoints with an arrow over top.

Example: $\overrightarrow{CD}$

Label each ray.

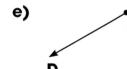  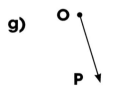 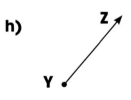

e) _____ f) _____ g) _____ h) _____

Identify each one as a line, line segment, or ray.

i) _____ j) _____ k) _____ l) _____

© CLASSROOM COMPLETE PRESS

17

Geometry CC3114

NAME: _____

Task Sheet 12

Volume of a Cube or Rectangular Prism

12) To find the volume of a cube or rectangular prism, multiply the length by its width and by its height.

Volume = length x width x height

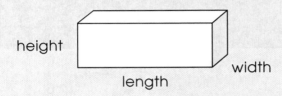

Find the volume of the cubes and rectangular prisms below.

a)

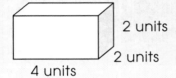

2 units
2 units
4 units

_____ cubic units

b)

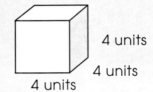

4 units
4 units
4 units

_____ cubic units

c)

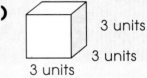

3 units
3 units
3 units

_____ cubic units

d)

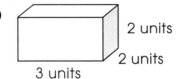

2 units
2 units
3 units

_____ cubic units

e)

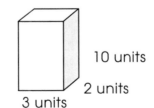

10 units
2 units
3 units

_____ cubic units

f)

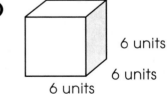

6 units
6 units
6 units

_____ cubic units

g)

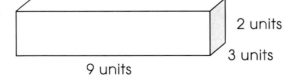

2 units
3 units
9 units

_____ cubic units

h)

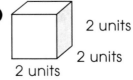

2 units
2 units
2 units

_____ cubic units

Geometry CC3114

NAME: _____

Task Sheet 13

Surface Area of a Sphere

13) To find the surface area of a sphere, square the radius and multiply by 4 Pi. (Pi = 3.14)

Surface area = $4\pi r^2$

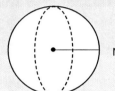 radius

Find the surface area of the spheres below.

a)

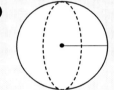 3.2 units

_____ square units

b)

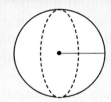 1.5 units

_____ square units

c)

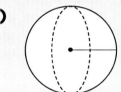 4 units

_____ square units

d)

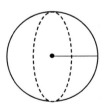 4.1 units

_____ square units

e)

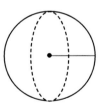 2.3 units

_____ square units

f)

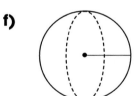 .6 units

_____ square units

Explore With Technology   Use the Internet to find out more about Pi.

Task Sheet 14

Surface Area of a Cylinder

14) To find the surface area of a cylinder, first find the area of the circles at the top and the bottom of the cylinder, then find the area of the middle part of the cylinder.

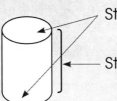

Step 1: Find the surface area of both circles.

Surface area = $2\pi r^2$

Step 2: Find the area of the middle of the cylinder.

Surface area of the middle = $2\pi rh$

Step 3: Add both surface areas together to get the total surface area of the cylinder.

Find the surface area of each cylinder below.

a)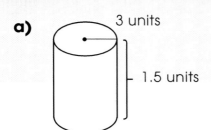

3 units

1.5 units

_____ square units

b)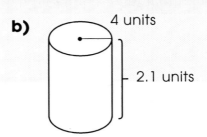

4 units

2.1 units

_____ square units

c)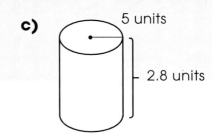

5 units

2.8 units

_____ square units

d)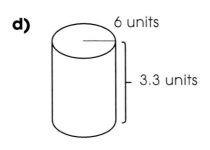

6 units

3.3 units

_____ square units

e)

2 units

4.2 units

_____ square units

f)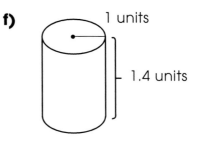

1 units

1.4 units

_____ square units

Task Sheet 15

Surface Area of a Cube and a Rectangular Prism

15) To find the surface area of a cube, square the length of one side, then multiply by 6. **Surface area = 6a²**

a

Find the surface area of the cubes below.

a)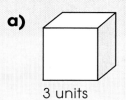
3 units

_____ square units

b)
2.5 units

_____ square units

c)
4.2 units

_____ square units

d)
5.1 units

_____ square units

To find the surface area of a rectangular prism, find the area of each side of the prism and multiply this number by 2. **Surface area = 2(ab) + 2(bc) + 2(ca)**

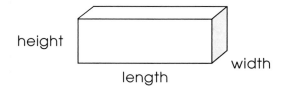
height
length
width

Find the surface area of the rectangular prisms below.

e)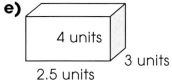
4 units
3 units
2.5 units

_____ square units

f)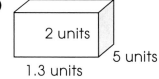
2 units
5 units
1.3 units

_____ square units

g)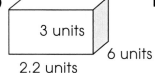
3 units
6 units
2.2 units

_____ square units

h)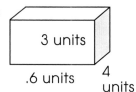
3 units
4 units
.6 units

_____ square units

Geometry CC3114

Drill Sheet 1

Draw each shape.

a) **Parallelogram**   b) **Trapezoid**   c) **Rhombus**

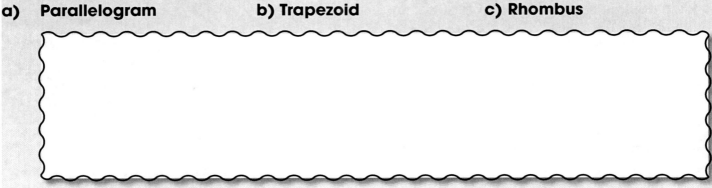

Measure each angle.

d)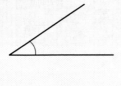

_____ °

e)

_____ °

f)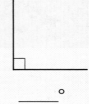

_____ °

g) **Find the area of the square.**

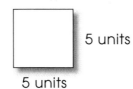
5 units
5 units

_____ square units

h) **Find the area of the rectangle.**

6 units
8 units

_____ square units

i) **Find the radius of the circle below.**

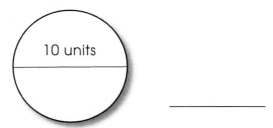
10 units

_____

NAME: _____

Drill Sheet 2

Find the area of each triangle below.

a)

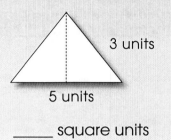

3 units

5 units

_____ square units

b)

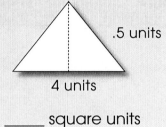

.5 units

4 units

_____ square units

c)

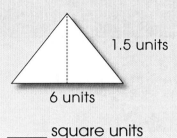

1.5 units

6 units

_____ square units

Use the circle to answer the questions.

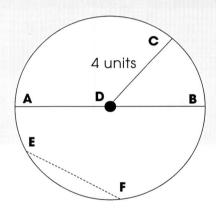

4 units

d) Which line segment is the diameter? _____

e) Which line segment is the chord? _____

f) Which line segment is the radius? _____

g) What is the area of the circle? _____

h) What is the diameter of the circle? _____

Use the shape below to answer the questions.

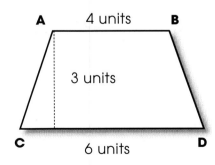

A   4 units   B

3 units

C   6 units   D

i) What is the name of this shape? _____

j) Which two lines are parallel? _____

k) What is the height of the trapezoid? _____

l) What is the area of the trapezoid? _____

Review A

a)   Draw each angle.

   i) 90°   **ii) 35°**   **iii) 150°**

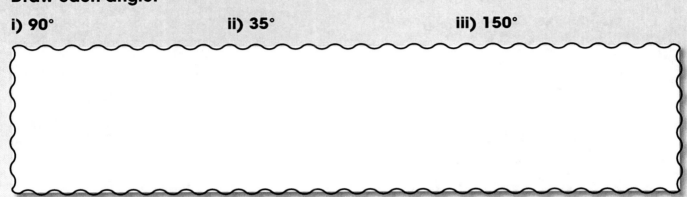

b)   What is an acute angle?   _____

c)   What is a right angle?   _____

d)   What is an obtuse angle?   _____

e)   Label all the right angle(s) in each shape below.

   i)   ii)    iii)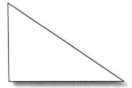

f)   Find the area of the parallelogram below.

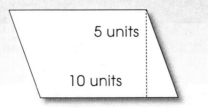

   5 units

   10 units   _____ square units

g)   Name the two ways of describing a triangle.

_____

_____

NAME: _____

Review B

a) **Find the missing angle.**

i)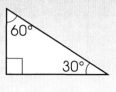

ii)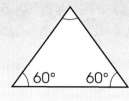

iii)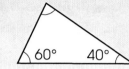

_____ °   _____ °   _____ °

b) **What are the differences between equilateral, isosceles, and scalene triangles?**

_____

_____

c) **What are the differences between right, obtuse, and acute triangles?**

_____

_____

d) **What is the formula for finding the area of a triangle?** _____

e) **Find the area of the following triangles below.**

i)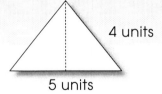

ii)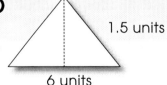

iii)

_____ square units   _____ square units   _____ square units

Geometry CC3114

Review C

a) **Draw the following types of lines.**

    i) Perpendicular   **ii) Parallel**   **iii) Intersecting**   **iv) Skew lines**

b) **Answer the questions using the lines below.**

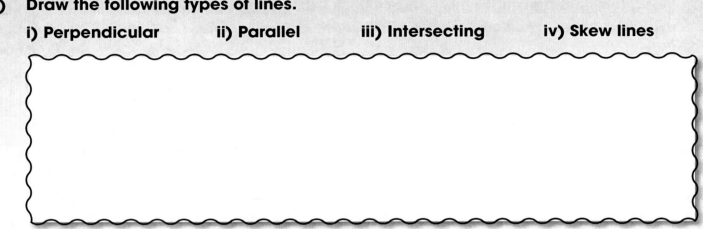

  i)   Which line is a line segment? _____

  ii)   Which line is a line? _____

  iii)   Which line is a ray? _____

c) **What is the formula for finding the volume of a rectangular prism or cube?**

_____

d) **Find the volume for the rectangular prism below.**

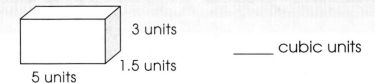

3 units

1.5 units

5 units

_____ cubic units

7.

a) Circumference

b) Chord

c) Radius

d) Diameter

e) radius = 1, diameter = 2

f) radius = 1.5, diameter = 3

g) radius = 2.2, diameter = 4.4

h) radius = 5.5, diameter = 11

(13)

6.

a) 15 square units

b) 22.5 square units

c) 16.5 square units

d) 21 square units

e) 21 square units

f) 10 square units

g) 6 square units

h) 20 square units

i) 7.5 square units

(12)

4.

a) 4.5 square units
b) 8 square units
c) 15.5 square units
d) 25.2 square units
e) 7.2 square units
f) 1.1 square units
g) d) with 25.2 square units
h) f) with 1.1 square units

(10)

5.

a) isosceles, right, 90 degrees

b) isosceles, right, 45 degrees

c) equilateral, acute, 60 degrees

d) scalene, acute, 85 degrees

e) scalene, obtuse, 115 degrees

f) isosceles, acute, 75 degrees

(11)

3.

a) 4 square units

b) 4.5 square units

c) 2.25 square units

d) 4.4 square units

e) 4.84 square units

f) 10.89 square units

g) 26.01 square units

h) 6.75 square units

i) 20.5 square units

(9)

2.

a) $\angle$ CDA = 105°

b) $\angle$ ABC = 90°

c) $\angle$ DAB = 105°

d) $\angle$ BCD = 90°

e) $\angle$ CDA = 70°

f) $\angle$ DAB = 90°

g) $\angle$ CBA = 90°

h) $\angle$ CDA = 120°

Angle DAB of quadrilateral f), and all angles of quadrilateral b), d), and g) are 90 degrees.

(8)

1.

a) 20 degrees, acute

b) 135 degrees, obtuse

c) 90 degrees, right

d) 60 degrees, acute

e) 90 degrees, right

f) 130 degrees, obtuse

g) 45 degrees, acute

h) 100 degrees, obtuse

i) 160 degrees, obtuse

j) 90 degrees, right

k) 110 degrees, obtuse

l) 30 degrees, acute

(7)

EZ✓

14.

a) 84.78 square units

b) 153.23 square units

c) 244.92 square units

d) 350.42 square units

e) 77.87 square units

f) 15.07 square units

20

13.

a) 128.61 square units

b) 28.26 square units

c) 200.96 square units

d) 211.13 square units

e) 66.44 square units

f) 4.52 square units

19

11.

a) $\overline{AB}$

b) $\overline{FG}$

c) $\overrightarrow{ST}$

d) $\overleftrightarrow{WX}$

e) $\overleftrightarrow{CD}$

f) $\overrightarrow{LM}$

g) $\overrightarrow{OP}$

h) $\overrightarrow{YZ}$

i) Line

j) Ray

k) Line segment

l) Line

17

12.

a) 16 cubic units

b) 64 cubic units

c) 27 cubic units

d) 12 cubic units

e) 60 cubic units

f) 216 cubic units

g) 54 cubic units

h) 8 cubic units

18

10.

a) Parallel

b) Intersecting

c) Skew

d) Perpendicular

e) Intersecting

f) Perpendicular

g) Parallel

h) Skew

i-l) Check to make sure the student drew the appropriate lines.

16

9.

a) 24 square units

b) 14 square units

c) 12 square units

d) 24 square units

e) 32 square units

f) 8 square units

g) 12 square units

h) 18 square units

i) 50 square units

15

8.

a) 153.86 square units

b) 28.26 square units

c) 7.07 square units

d) 63.59 square units

e) 200.96 square units

f) 50.24 square units

14

Review C

a) Check to make sure the student drew the appropriate lines.

b)
i) $\overline{EF}$
ii) $\overline{CD}$
iii) $\overrightarrow{AB}$

c) length x width x height

d) 22.5 cubic units

26

Review B

a) i) 90 degrees
ii) 60 degrees
iii) 80 degrees

b) The difference is the number of equal sides. An equilateral triangle has three sides of the same length. An isosceles triangle has two sides of the same length. A scalene triangle has no sides of the same length.

c) A right triangle has one angle of 90 degrees. An obtuse triangle has an angle greater than 90 degrees. An acute triangle has all three angles less than 90 degrees.

d) base x height divided by 2

e) i) 10 square units
ii) 4.5 square units
iii) 7.5 square units

25

Review A

a) Check to make sure the student drew the appropriate angles.

b) Any angle less than 90 degrees.

c) An angle of 90 degrees.

d) Any angle greater than 90 degrees.

e) i) Each corner of the square is a 90° angle.
ii) There are no 90° angles.
iii) The corner angle is 90 degrees.

f) 50 square units

g) Describe a triangle by the type of angles and number of sides with the same lengths.

24

Drill Sheet 2

a) 7.5 square units

b) 1 square unit

c) 4.5 square units

d) $\overline{AB}$

e) $\overline{EF}$

f) $\overline{CD}$

g) 50.24 square units

h) 8 units

i) Trapezoid

j) $\overline{AB}$ and $\overline{CD}$

k) 3 units

l) 15 square units

23

Drill Sheet 1

a) - c) Check to make sure the student made the appropriate shape.

d) 35 degrees

e) 50 degrees

f) 90 degrees

g) 25 square units

h) 48 square units

i) 5 units

22

15.

a) 54 square units

b) 37.5 square units

c) 105.84 square units

d) 156.06 square units

e) 59 square units

f) 38.2 square units

g) 75.6 square units

h) 32.4 square units

21

EZ✔

6.

a) i) Rectangle

ii) Parallelogram

iii) Trapezoid

iv) Hexagon

v) Square

vi) Oval

vii) Pentagon

viii) Rhombus

ix) Right triangle

b) i) Rectangle, v) Square, viii) Rhombus, ix) Right triangle

c) i) Rectangle, iii) Trapezoid, iv) Hexagon, v) Square, vi) Oval, vii) Pentagon, viii) Rhombus

d) vi) Oval

6A

5.

a)
Check to make sure the student drew the appropriate angles.

b) i) 45°, iii) 10°, iv) 65°, vii) 30°

c) ii) 115°, vi) 135°, viii) 100°, ix) 180°

d) ix) 180°

5A

4.

a) 1

b) 0

c) 2

d) 1

e) 4

f) 0

g) 4

h) 1

i) 2

j) 0

k) 3

l) 0

m) 0

n) infinite

o) 1

p) 8

4A

3.

Answers may vary. Possible answers include:

l = 4, w = 3, h = 2;
l = 8, w = 3, h = 1;
l = 12, w = 2, h = 1;
l = 6, w = 2, h = 2;
l = 6, w = 4, h = 1

The volume of each rectangular prism will be 24 cubic units.

3A

2.

a) v

b) vi

c) iv

d) i

e) iii

f) ii

2A

1.

Answers may vary. Possible answers include: parallelogram and triangle.

1A